LEAN CONSTRUCTION BASICS

Table of content

1	INTRODUCTION
7	5S
19	KANBAN
29	ADDED VALUE
40	LAST PLANNER SYSTEM
60	ZONING
66	PRODUCTION PROCESSES
77	ADMINISTRATIVE PROCESSES
87	CHANGE MANAGEMENT
98	OFF-SITE
104	LINK WITH THE BIM
115	CONCLUSION

AF491019

0

INTRODUCTION

INTRODUCTION

Lean is based on the idea of reducing everything unnecessary ; the actions you don't need to get the work done.

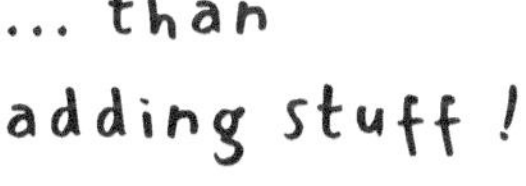

Lean is directly related to <u>continuous improvement</u> methods, which are based on a way of thinking about our working methods at all times.

INTRODUCTION

Lean Management appeared in the car industry, especially in Toyota in the 1950s.

When they hired a new ingeneer, he was given an area in the factory ; in this very perimeter, he had to analyze everything that was going on.

The job of the ingeneer was, for the next years, to improve the production of this area, and only this one.

Lean was born.

INTRODUCTION

A Lean method is based on 4 pillars :

- Cost reduction
- Reduction of lead times
- Quality improvement
- Increased safety

As a bonus, we can notice a global peace of mind in a company with a Lean organisation, and this is quite a key point nowadays.

INTRODUCTION

The expression "Lean Construction" was first prononced in the USA in 1990; decades later, it's been applied in many countries and keeps developing.

"We only produce prototypes". First of all, this is questionable : look at buildings around you. Then, Lean Constructions aims at focusing on the construction <u>mode</u>, not the product.

The idea is to analyze and improve the <u>processes</u> which lead to those products.

INTRODUCTION

Lean is a mix between :
- A toolbox.
- A formation work along with change management.

The applications are multiple: on the construction site, but also internally in your company.

To sum up, Lean Construction is by 80% Lean Management, with a few specific tools developed for construction, such as Last Planner System®.

TOOLBOX
5S

TOOLBOX
5S

5S aims at creating tidy and optimized workspaces according to their use.

Easy on the paper, right ?

This actually represents a perfect starting point in Lean: The results are pretty quick and, more important, everyone can see it: workers are softly dragged in a mindset change unconsciously.

TOOLBOX
5S

5S comes from Japan. It has been traduced in english like this:

- Sort
- Set in order
- Shine
- Standardise
- Sustain

The continuous application of those 5 principles is the basis of the 5S approach.

TOOLBOX
5S

To maximise the chances of succeed in the 5S implementation, the workers must be made aware of the continuous improvement basics ; if they don't involve themshelves in the process, this can't work.

This encourage their commitment to the project. For each case, the user must take part in the branstorming and design sessions.

TOOLBOX
5S

Current situation : the warehouse is disturbed, disorganized. Workers spend half an hour each morning and half an hour each evening looking for items.

Application of 5S : the warehouse is tidy and organized. The workers now spend only 15 minutes in the morning and 15 minutes in the evening looking for items.

Each worker saves half an hour a day, or 10 hours a month, or 120 hours a year : 3 full weeks of production per worker are saved every year! I let you do the math to convert in your country's hourly rate... Pretty cool, Huh ?

TOOLBOX
5S

What are the benefits on the construction site?

Without the maths, here are the observable benefits on site:

- Dedicated storage space
- Reduced handling time
- Reduction of conflicts
- Increased security

TOOLBOX
5S

Still without calculation, here are the observable benefits of 5S in an office :

- Various losses (documents...) reduced
- Clarity of space use
- Increased productivity
- Increased serenity

TOOLBOX
5S

For this part, we will apply 5S on construction sites and depots.

SITES	- No "unknown" items on site - Any out-of-area items must be treated
DEPOT	- Empty all the elements and think about their organization - Keep only what is useful (not "what could be useful")

TOOLBOX
5S

How to manage the storage part in a building site and a warehouse?

SITE	- Everyone is responsible for their own storage - ... but also for the whole common area !
DEPOT	- Reserve space for stored references - Create adapted storage methods (racks, cupboards, boxes...)

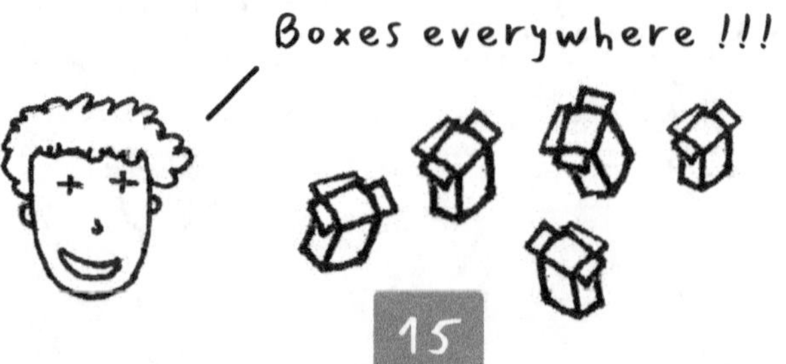

TOOLBOX
5S

How to manage the cleaning part in a construction site and a warehouse?

SITE	- Every day, the workers stop working 10 minutes before the end of the day to clean.
DEPOT	- Cleaning routine to be put in place, with a responsible person.

TOOLBOX
5S

How to manage the standardization part in a construction site and a warehouse?

SITE	- Mark out areas and paths - Indicate the nature of the area (signs, markings...)
DEPOT	- Label storage areas - Draw the zones - Display standards (picture of the row zone in front of the zone)

TOOLBOX
5S

How to manage the sustainability part in a construction site and a warehouse?

SITE	- A third party performs controls of the 4 previous criteria while checking the progression
DEPOT	- Regular control of the criteria - Measurement of the reduction of out-of-stock situations

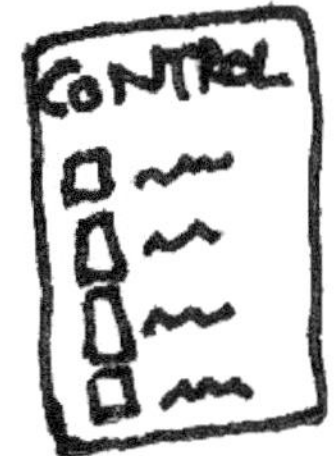

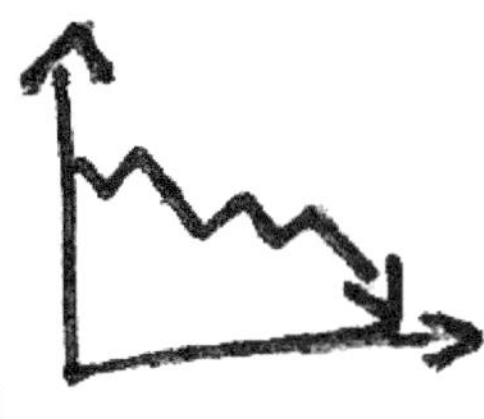

TOOLBOX KANBAN

TOOLBOX
KANBAN

Kanban is - again - a tool from the industry.

Here is the overall principle :

An assembler receives an order for a table. He sends a "kanban" sheet to each workshop to inform them of the parts needed from their workshop.

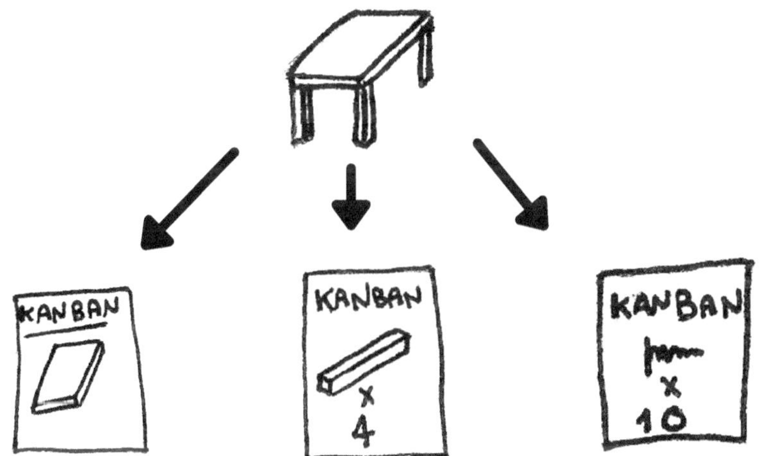

When these parts are returned to the assembler (along with the sheets), the assembler can link the parts and fulfill the order.

TOOLBOX
KANBAN

The use of these cards is intended to facilitate communication between different departments and workers in a company.

The location of the product sheet gives information on its status.

Let's take the previous example:

- If a kanban sheet is in a workshop, the workshop must create the linked product.

- If a kanban sheet is back at the assembler, then the product is made and ready to be assembled.

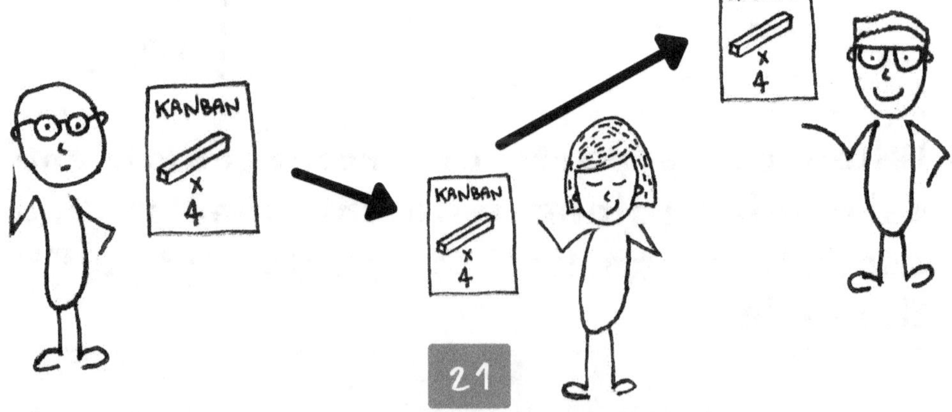

TOOLBOX
KANBAN

The Kanban method is applicable in construction companies, in particular to manage their stocks.

Inventory is often a expensive item in a company; reducing it and eliminating stock shortages automatically generates cash.

The objective is to obtain an efficient, economical and sustainable inventory management.

Applications can also be made on some large sites, in addition to 5S.

TOOLBOX
KANBAN

How to get started?

An inventory of existing systems is always the best place to start:

- Who places the orders?
- Who reports lack of stock?
- Who stores the orders?
- Who controls the deliveries?
- Etc.

The objective is to obtain as much information as possible about the existing system, in order to adapt the tools as well as possible.

TOOLBOX
KANBAN

How to limit out-of-stock situations?

First, identify the items belonging to the "current stock":

- They are ordered regularly
- They are consumed regularly

For each of these elements, the objective is to determine a safety stock (also called "buffer stock").

Current stock

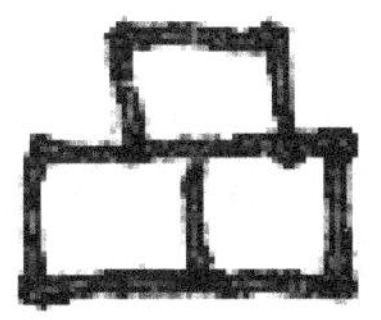

Buffer stock

This stock represents the quantity that will be consumed before the next delivery: this way, stockouts are avoided!

TOOLBOX
KANBAN

In storage, the kanban sheet must be placed in front of the buffer stock.

This means that to reach the buffer stock, it will be required to remove this sheet.

Thus, the sheet must be adapted in terms of size, but also in terms of system to match your needs! You can find examples on the internet for inspiration, but the best system will be the one adapted to your stock.

TOOLBOX
KANBAN

The technicians consume the stock until they "hit" the kanban sheet.

As explained above, they are forced to remove the card to start the buffer stock: they have to collect it.

Finally, they move it to a location intended to trigger the order, before they start the buffer stock.

TOOLBOX
KANBAN

How to restock?

Once or several times a week (depending on your needs), the person in charge of ordering will pick up the sheets from the place where the technicians dropped them off.

The person can order the missing references in the warehouse (it is useful to write the quantities to be ordered directly on the sheets).

At delivery, a responsible person collects the sheets and puts the stocks back in place (with their kanban sheet).

TOOLBOX
KANBAN

The results of this type of approach are the following ones :

- Less (or no) stock-outs

- Regulated, efficient communication, without "blind spots"

- Employee peace of mind (no need to alert if a reference is missing, the system works by itself)

TOOLBOX
ADDED VALUE

TOOLBOX
ADDED VALUE

What is added value?

Added Value refers to everything the customer pays for.

What does it mean ? Your customer doesn't pay for your waste, no matter how much of it you produce.

TOOLBOX
ADDED VALUE

For example, a customer pays for the pouring of the wall (finished product), but not for the time it takes to install the panels (at least not directly).

 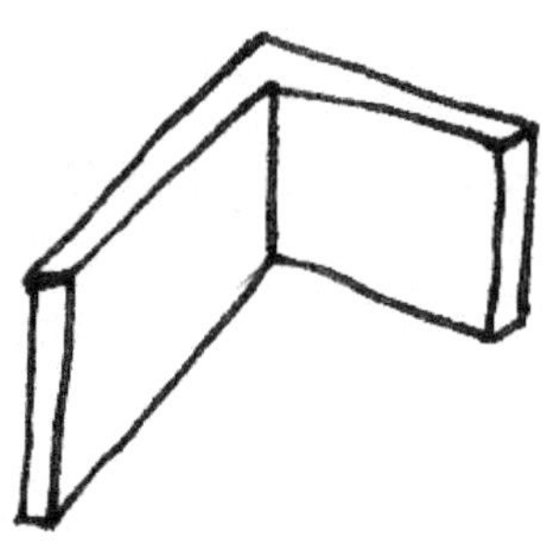

Another example: the customer pays for the painting of the wall (added value), but not for your return trip to the supplier because you forgot a brush (waste).

TOOLBOX
ADDED VALUE

This is of course the case. That's why there is a third category:

The Value Added Process.

Like waste, the customer does not pay for this category, but it remains necessary to create added value.

For example, fixing a support to core a wall is Value Added Process. The customer does not pay to fix the support, but it must be done to generate Added Value.

TOOLBOX
ADDED VALUE

Today, in the building industry, the ratios are indicative of the progress that can be made:

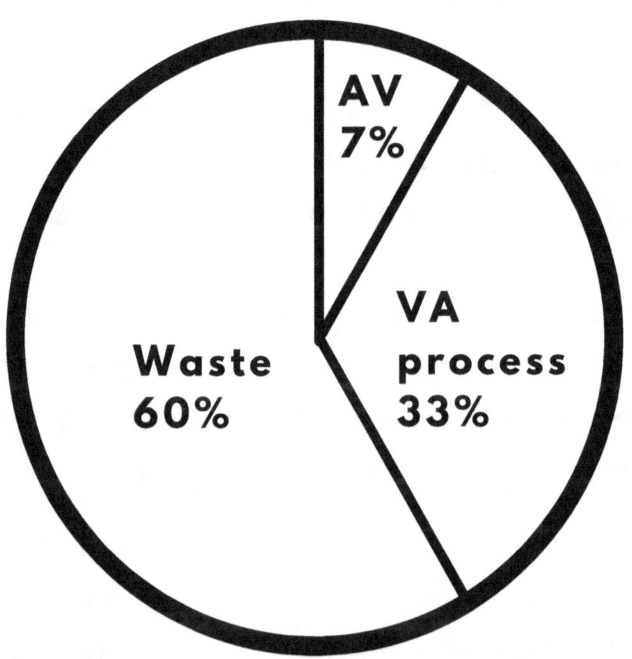

Source : DeltaPartners.fr

You got it, the objective is to reduce the waste, which represents 60% of the resources used on site.

TOOLBOX
ADDED VALUE

Let's take a basic example to explain this phenomenon:
Sawing a board.

Added Value:
Sawing to create the finished product, which is the board cut in half.

Value Added Process:
The blocking of the board and the time to get the tool.

Waste :
The board badly blocked so badly sawn: need to start again, and the search for saw in a chaotic depot.

TOOLBOX
ADDED VALUE

After measurement, the following results are obtained:

	Time (s)	Designation
W	600	Tool searching
VA PRO	60	Bringing tool
W	60	Board blocking
W	60	Board sawing
W	120	Badly blocked board Search for another board
VA PRO	60	Blocking new board
AV	60	Sawing new board

Summary : 1 minute of AV, 2 minutes of VA Process and 14 minutes of Waste.

Total time : 17 minutes

TOOLBOX
ADDED VALUE

How can this be improved?

The objective is to reduce the total execution time, and we notice right away that the biggest item is the waste (14 minutes).

Here are some examples to reduce it:

- Tool search :

> By applying a 5S approach to optimize the workspace, this search does not take place: 10 minutes saved!

- First sawing with defect

> By focusing on quality, not speed, and therefore taking your time to block the board, you only have to execute the process once.

TOOLBOX
ADDED VALUE

After applying these principles, we obtain the following measurements:

	Time (s)	Designation
VA PRO	60	**Bring tool**
VA PRO	180	**Board blocking**
AV	60	**Sawing of the board**

The **NVA** is completely removed, and time is allocated to the **VA** process : the blocking of the board goes from one to three minutes.

Result: the total time is only 5 minutes! This saves 12 minutes of execution time for each board you need to saw...

TOOLBOX
ADDED VALUE

We notice that this method applied on a larger scale can generate significant time and productivity gains... But be careful !

This is only effective after objectively measuring the facts.

Only with a timer and by analyzing all the actions performed is a waste reduction possible.

TOOLBOX
ADDED VALUE

On a construction site, the biggest waste "item" is often the workers' walk: all these useless trips can generate hidden costs and colossal losses at the end of the year.

As a measurement method, a pedometer allows to measure the progress made during the implementation of corrective actions.

The fact of thinking ahead about the workspace on the site reduces travel: this approach can of course be coupled with the 5S.

TOOLBOX

LAST PLANNER SYSTEM

TOOLBOX
LAST PLANNER SYSTEM

We are going to focus on a purely "construction site" tool, the Last Planner System. Here are its objectives:

- Reduce construction time.

- Reduce costs.

- Reduce conflicts and friction between companies on site.

- Anticipate and deal with various problems.

TOOLBOX
LAST PLANNER SYSTEM

This is a collaborative planning for construction sites.

All the actors of a construction site are involved in its planning.

No more Gantt schedules imposed by a project manager you've never seen!

Be careful: this is a complete approach, not a magic tool !

TOOLBOX
LAST PLANNER SYSTEM

This approach can be broken down into four phases with different tools.

Do not hesitate to adapt the tools to the specificities of your project: this is the essence of continuous improvement.

All the actors must be trained to make it work... One disrupter is enough to break the whole system.

TOOLBOX

LAST PLANNER SYSTEM

MASTER SCHEDULE

TOOLBOX
LAST PLANNER SYSTEM

The first draft of the planning to be done; you have to start somewhere!

Specifications :

- General planning, global vision

- Generic lots only (no macro-lots or sub-lots)

- Coarse input data (milestones, key dates...)

- Realized by the supervisory team

- Type: Gantt

TOOLBOX
LAST PLANNER SYSTEM

The Master schedule is built in three very simple steps, with a classic planning method:

- Create a time chain (on a spreadsheet for example)

- List the companies that must intervene

- Represent the intervention times

TOOLBOX
LAST PLANNER SYSTEM

The master schedule aims at:

- Visualizing the major ensembles

- Visualizing the main actors

- Placing the project in time

This is not the finished plan! It is only the starting point that will form the basis for everything else.

TOOLBOX

LAST PLANNER SYSTEM

PHASE SCHEDULING

TOOLBOX
LAST PLANNER SYSTEM

Phase scheduling is the second tool in the process.
It is a key tool in the Last Planner Sytem.

Specifications :

- Multi-month vision

- Level of detail : week

- Completed as soon as work has begun

- Collaborative

TOOLBOX
LAST PLANNER SYSTEM

The timeline is created on a whiteboard in a defined room.

Each company gets a color of post-it notes.

On each post-it note, the company writes:

- The name of the task

- The execution constraints

- The prerequisites of the task

Then, each company places its interventions on the schedule (in principle, one post-it = one week)

TOOLBOX
LAST PLANNER SYSTEM

The starting point for the construction of this schedule is the finished product (what we want to deliver).

Everyone sets up their interventions and checks that the interactions with the other professionals are fluid: this allows to remove problems of overlapping before the execution.

The planning is optimized and the companies work together: this is the essence of the Last Planner System.

TOOLBOX

LAST PLANNER SYSTEM

LOOK-AHEAD PLANNING

TOOLBOX
LAST PLANNER SYSTEM

It means "Look into the future". Complementary with the previous resources, this one allows an automatic follow-up of the good realization conditions of the work.

Specifications :

- Progress schedule over 6 to 8 weeks

- Weekly follow-up

- Verification of the achievement of the prerequisites

- Binary follow-up (Yes / No)

TOOLBOX
LAST PLANNER SYSTEM

How to build the "Look-ahead planning"?

The look-ahead planning is presented in the form of a collaborative table.

In this table, all the potential blocking points are listed for each task of a project.

In case of missing items, the table sends an alert or visually indicates that prerequisites are missing.

TOOLBOX
LAST PLANNER SYSTEM

In use, look-ahead planning acts on two important points:

- It is always cheaper to correct a mistake on the plans than to break a wall!

- The actors of the construction site are placed in front of their responsibilities; the sharing of information avoids transmission errors, which are very frequent with e-mails today.

TOOLBOX

LAST PLANNER SYSTEM

PRODUCTION PLANNING

TOOLBOX
LAST PLANNER SYSTEM

To conclude this chapter on LPS, here is one last tool : production planning.

It is the tool that will allow to fluidify the day to day interventions on site.

Specifications :

- Weekly meeting on a fixed day

- Vision : one to two weeks

- Each meeting has three phases

TOOLBOX
LAST PLANNER SYSTEM

Phase 1 of the meeting aims at analysing the past week: without measurement, we cannot progress!

Thus, the following elements are measured:

- What work has been done

- What work has not been done

- The difference between the planned work and the completed work (Percentage of Promises Achieved - PPC)

Then, a problem solving is carried out in order to understand the gaps (5 Whys method, 8D)

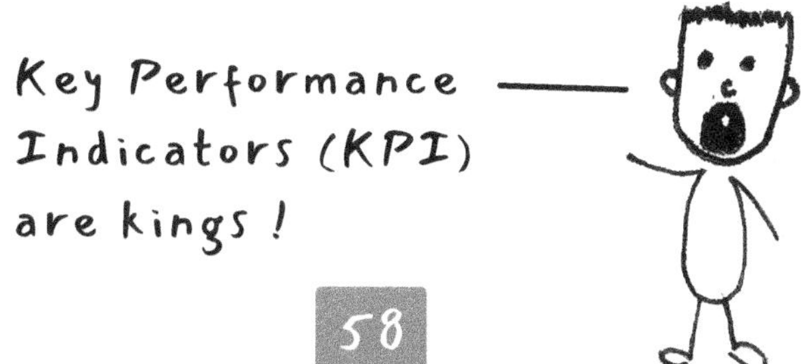

TOOLBOX
LAST PLANNER SYSTEM

Phase 2 of the meeting is about planning the following week. Everyone announces the work they <u>can</u> do in the coming days.

The data from this phase 2 will then be used for phase 1 of the following week.

Finally, phase 3 allows us to take a step back from the next few weeks of work, and to anticipate the blocking points in the longer term.

TOOLBOX
ZONING

TOOLBOX
ZONING

What does that mean?

Let's take the traditional planning of a building site: In many projets, only one building trade works at the same time on site.
The mason does the job, then the finishing companies follow one another from building to building, from floor to floor...

The planning with the zoning consists in basing itself on the zones of the building site, and not the trades; and that makes all the difference.

TOOLBOX
ZONING

What is the initial problem?

The many delays on construction sites today are quite misunderstood; at the end, you don't really know where the lost time went (and no, it's not always the fault of that company you don't like).

We can then ask ourselves: how to shorten the deadlines without reducing the execution time of each one?

The solution proposed by Zoning is the following one: find a way for everyone to intervene as soon as possible.

TOOLBOX
ZONING

As previously explained, zoning consists in delimiting zones in order to build a planning on the basis of a spatial division.

A principle to deeply understand is that the smaller the division (the smaller the areas chosen), the more time is saved: it's mathematical!

TOOLBOX
ZONING

Let's take the floor of a typical construction site.

The tiler, the electrician, the plumber and the plasterer must now pass one after the other to complete their work.

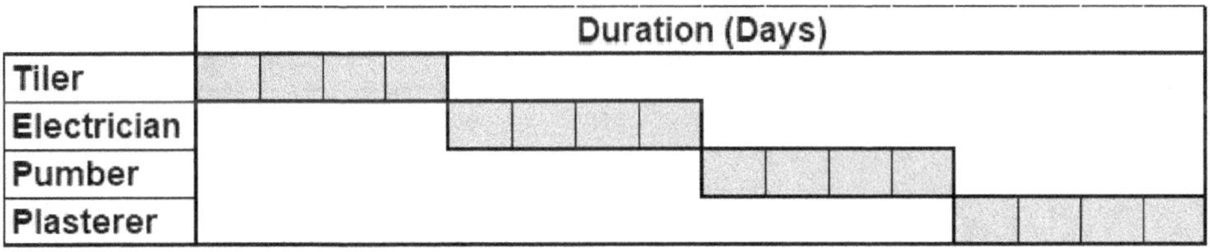

	Duration (Days)		
Tiler	▓▓▓▓		
Electrician		▓▓▓▓	
Pumber			▓▓▓▓
Plasterer			▓▓▓▓

The duration of the work is 16 days.

TOOLBOX
ZONING

For this case, we observe that it is possible to divide into four zones of 400m^2 each:

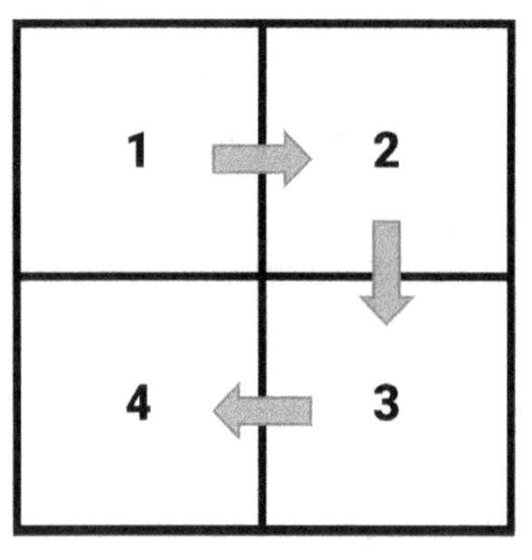

400 m² areas

Each craftsman works from zones 1 to 4, and each one follows in the wake of the previous one.

	Duration (Days)						
Tiler	1	2	3	4			
Electrician		1	2	3	4		
Pumber			1	2	3	4	
Plasterer				1	2	3	4

The duration of the worksite is now only 7 days, i.e. 56% shorter! And this, without reducing intervention times.

TOOLBOX

PRODUCTION

PROCESSES

TOOLBOX
PRODUCTION PROCESSES

In these parts, we will look at Lean from inside the construction companies. How to implement Lean in my organization?

Each company is characterized by a set of processes.

In the case of production, the processes represent all the actions to be carried out to deliver a finished product.

TOOLBOX
PRODUCTION PROCESSES

Identifying, locating and optimizing critical processes will allow you to:

- Save time

- Reduce parasite communication

- Eliminate unnecessary actions

- Save money

- Gain in serenity

- Have a global vision of your production

TOOLBOX
PRODUCTION PROCESSES

How to identify your processes?

First of all, as in any Lean approach, some preliminary measures are necessary. Here is how to proceed:

- List the finished products (toilet, electrical outlet, baseboard, etc)

- Bring together the production actors

- Based on the finished product, build the chain of actions that leads to the production of the deliverable.

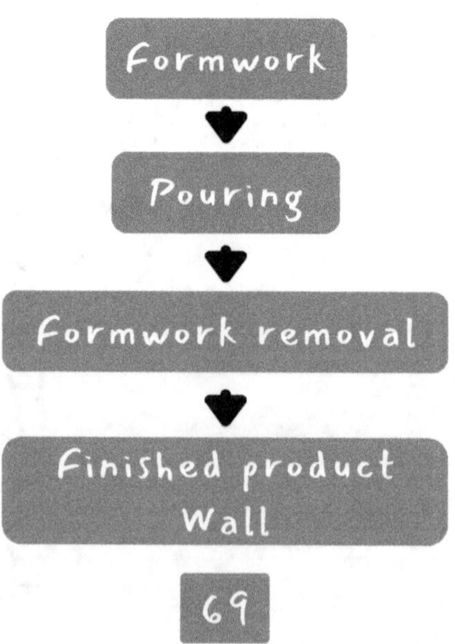

TOOLBOX
PRODUCTION PROCESSES

What are the elements to be identified for each process?

Each process must be represented as completely as possible: from the ordering of the material to the installation on site.

On each process represented, must appear:

- Who performs each step.

- The prerequisites of each step.

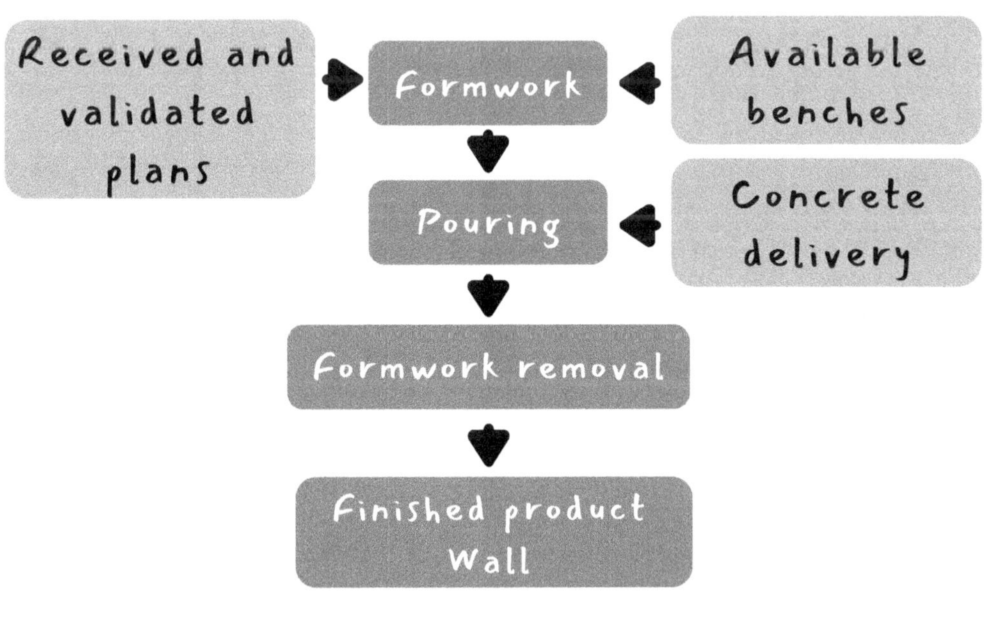

TOOLBOX
PRODUCTION PROCESSES

The steps of identification and representation of the processes make it possible to identify the most critical ones, i.e. those which present these characteristics:

- More time and energy invested

- Numerous difficulties encountered, generating time losses

- Variable execution time: the process is not defined / badly managed.

TOOLBOX
PRODUCTION PROCESSES

You thought it was over? No, there are still plenty of different reasons why a process might appear critical to your business.

Here are some others:

- The immutable nature: a process perceived as "fatal": you can't do otherwise...

- Source of financial losses

- Generator of stress at each execution

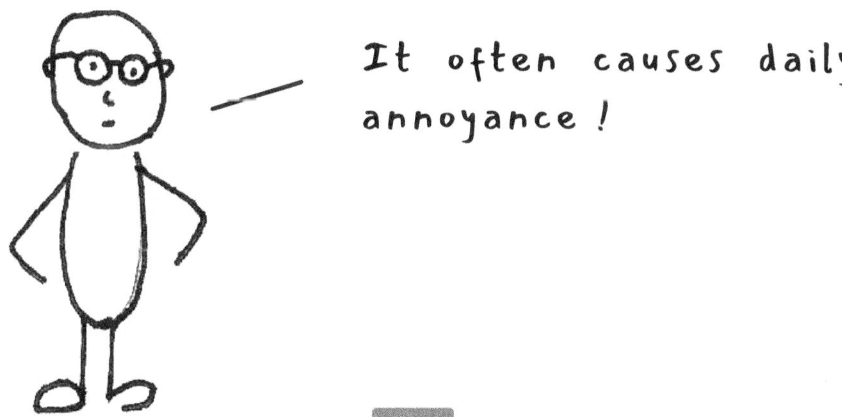

TOOLBOX
PRODUCTION PROCESSES

And above all, why would we represent them? After all, a quick chat at the 10 o'clock break is enough, right?

Of course not.

Effective representation is done in the following way:

- In a collaborative way: all actors of the process must validate their role.

- On a whiteboard, in a room dedicated to Lean.

TOOLBOX
PRODUCTION PROCESSES

"It's not rocket science to represent what's going on..."

But you have to do it to bring out the sources of improvement!

Here are the last elements to ensure an effective representation:

- Color-coded post-it friezes are very easy to read and understand.

- All elements, all constraints must appear, without exception.

TOOLBOX
PRODUCTION PROCESSES

This is what we are ultimately interested in; in fact, it is natural to go directly to this step without effectively representing it first.

This is one of the reasons why decisions are made without a global knowledge of the system and why the system gets stuck over time.

Here are some ways to optimize the system:

- The inventory carried out previously has probably already given rise to some thoughts

- The process must be fully questioned

- Identify actions that can be eliminated

TOOLBOX
PRODUCTION PROCESSES

How to optimize processes?

How to manage what can't be eliminated?

Several approaches are possible to further improve your processes:

- Identify the actions to be implemented to anticipate constraints

- Analyze work methods (observe innovation, new solutions)

- Focus on the most time-consuming tasks to try to reduce their duration

TOOLBOX

ADMINISTRATIVE PROCESSES

TOOLBOX
ADMINISTRATIVE PROCESSES

This is the global principle that will be repeated in order to work on the administrative processes.

The schemes to be applied are the same, and the benefits are similar:

- Time saving

- Elimination of hazardous actions

- Savings

- Serenity

- Etc.

TOOLBOX
ADMINISTRATIVE PROCESSES

Only the scope changes compared to the production processes.

Earlier, we took as an example of a finished product a toilet or an electrical outlet, while the finished products of the business processes are for example:

- A quote

- An invoice

- The placing of an order

- ... In short, all the actions of an office.

TOOLBOX
ADMINISTRATIVE PROCESSES

To understand and make an efficient inventory of your administrative processes, here are the two steps:

- Bring together all the actors who have to manage administrative documents: the management assistant, the secretary, the work manager, the accountant....

- Identify the needs of each one for each contract: estimates, invoices, various certificates... Who needs what?

TOOLBOX
ADMINISTRATIVE PROCESSES

Starting, as before, from the finished product, the objective is to identify the following elements:

- The list of prerequisites (for example, price information is a prerequisite for making a quote)

- The chain of actions performed today

- The links between all the actors

TOOLBOX
ADMINISTRATIVE PROCESSES

Even if the starting factors differ in relation to the production processes, the consequences that emerge converge:

- Generator of financial losses

- Generator of stress

- Excessive investment of time and energy

TOOLBOX
ADMINISTRATIVE PROCESSES

We can also mention these factors which, even if they are less measurable, remain good indicators of the criticality of a process:

- Sterile debates around the management of the process (on the administrative management of cases for example)

- Seen as if nothing could be done

- Regularly irritating

83

TOOLBOX
ADMINISTRATIVE PROCESSES

How to picture the processes?

But is this actually the same chapter twice?

Yes and no, the idea is to understand that process analysis, whatever its application, follows the same patterns.

To represent the process, it must be:

- In a collaborative way; this raises inconsistencies from the first representation

- Using color codes, signs, symbols...

- Representing all constraints

TOOLBOX
ADMINISTRATIVE PROCESSES

How to optimize administrative processes?

The optimization of every process starts with the rigorous inventory you have just made.

Then, here are some areas of improvement:

- Questioning the process completely; is the finished product useful, in the right form, etc.?

- Eliminate duplication

- Search for adapted tools

TOOLBOX
ADMINISTRATIVE PROCESSES

How to optimize administrative processes?

Observing the well-represented process can be enough to improve it without even following optimization methods.

But since you ask, here are other tracks to exploit:

- Questioning the legitimacy of each step (the number of useless actions performed in companies is very, very impressive).

- Re-assigning people to the right place; everyone can get closer to their expertise area.

CHANGE MANAGEMENT

CHANGE MANAGEMENT

That's right. We've always done it that way, after all!

The change in business and techniques pushes everyone to continuously reinvent themselves in order not to be overtaken.

Today, change is no longer an option to ensure the sustainability of your activities, which is why Lean methods, based on perpetual change, make sense.

CHANGE MANAGEMENT

How does change work?

When you change work methods or processes in an organization, the reactions are diverse.

However, there is one common reaction to change: we all resist it, and that's natural.

Here is the evolution of the effectiveness of employees exposed to change:

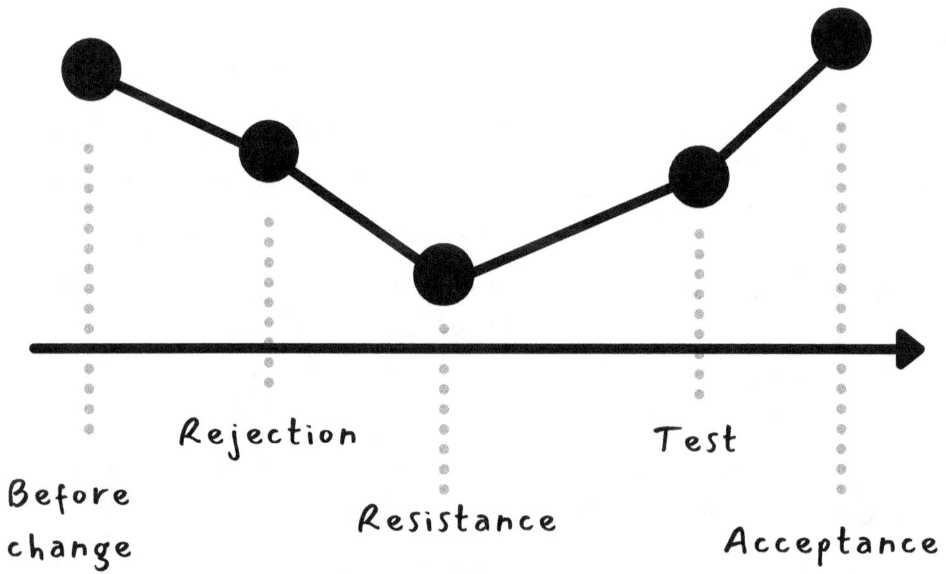

CHANGE MANAGEMENT

Even though the reactions you will get from your changes are not controllable, there are some principles to build on to make the process easier:

- Get adherence

- Communicate

- Coordinate teams

- Know how to manage time (plan)

- Be a "shrink", i.e. listen to everyone

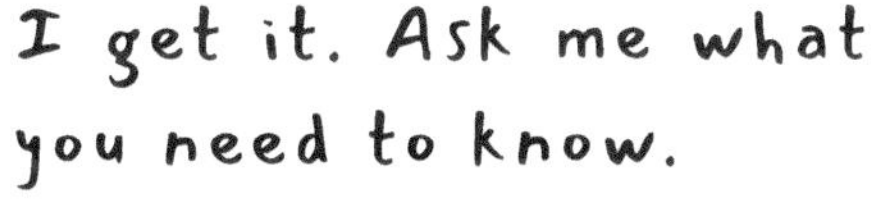

CHANGE MANAGEMENT

That's right. After many measurements, 4 main categories of attitudes to change emerge:

- Acceptance (about 20%)

- Indifference (about 60%)

- Passive resistance (about 15%)

- Active resistance (about 5%)

The objective is to create a driving team with the 20% of acceptance, and then to gradually bring all the groups on board.

CHANGE MANAGEMENT

This has also been studied. At the origins of Lean, it appeared that visual management is a very powerful tool to get ideas spread through the company.

The principle is simple: present data related to the change in a meeting room called "Obeya room", to encourage collaboration around the project.

CHANGE MANAGEMENT

The goal is to display the key information of your project, such as:

- The defined objectives

- 'To Do' tables to visualize the progress

- Before and after pictures, if you have done a 5S in the warehouse for example

- Progress indicators, allowing you to measure improvements (essential to understand the impact of a Lean approach!)

To communicate, we must also exchange?

In addition to visual management, it seems obvious that weekly (at least!) exchange routines should be set up to maintain the dynamic.

In these meetings :

- Everyone can express themselves (being directive kills creativity)

- An agenda is prepared: the objective is to be efficient and to deal with the issues quickly without drifting into the accountant's ski weekend.

CHANGE MANAGEMENT

These routines aim at keeping the actors involved in the continuous improvement project.

They are also characterized by :

- Detecting problems through the '5 Whys': going back to the source to continuously improve skills.

- Celebrating progress: I believe in your creativity for this part...

CHANGE MANAGEMENT

Your behavior as a driver of the process is crucial.

Here are some examples of good practices:

- Don't be judgmental: our cognitive biases make us perceive phenomena differently from reality. To keep your credibility intact, avoid "it will work... it won't work...". Just try it.

- Measurement is king: if you question the effectiveness of a particular position, measure its performance before you start anything.

CHANGE MANAGEMENT

As the notions of Lean and change are very unfamiliar to the construction industry, you will be the key contact for everyone in the company.

That's why you also need to:

- Communicate about the benefits for everyone (not just the overall goal, but the improvement of working conditions for example).

- Continuous support, understanding that changing habits is very difficult.

- Be aware that the lack of communication is definitely fatal.

OFF-SITE

OFF-SITE

And most importantly, what does this have to do with Lean Construction?

Off-site construction is based on the idea of assembling prefabricated blocks on site, while the manufacturing is done in the factory.

This principle is very representative of the progressive industrialization of the construction sector.

The application of Lean is like a "light" industrialization of some processes, when off-site pushes the phenomenon to the maximum.

OFF-SITE

Let's simplify the representation of the flows:

- In a traditional construction site, each truck of materials delivers its elements on site: the flows (and the uncontrolled consequences of their superposition) are very important.

- In an off-site construction site, a single truck delivers and installs the prefabricated elements on site.

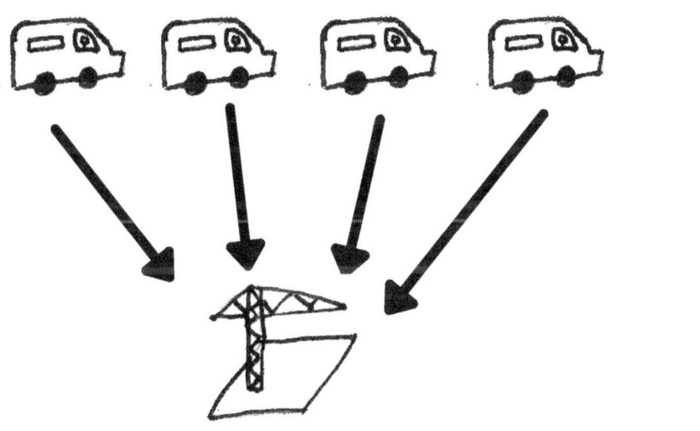

Classic construction site

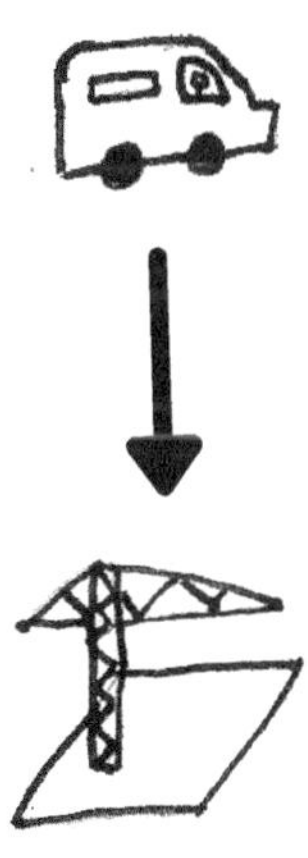

Off-site

OFF-SITE

In addition to the ease of construction site execution, off-site construction has many advantages:

- Optimal manufacturing conditions: no exposure to the elements in the factory (no rain, no snow, etc.)

- Speed, simplicity, savings

- Factory-like quality control: reduction or even elimination of defects

- Strong reduction of the carbon footprint

What is the interest compared to our current methods?

Factory-based manufacturing provides significant advantages over conventional construction:

- Computerized production management, to get closer to factory reliability

- Centralized management:
  - Supply
  - Quality control
  - Production monitoring
  - Product tracking
  - Workers management

OFF-SITE

Will our jobs disappear?

Without disappearing, they will definitely evolve: though, the required skills to assemble an electrical panel remain very similar on site or off site.

Yet, constructors will certainly become assemblers over time.

Moreover, less specialization is required for the workers thanks to the standardization of the processes: an interesting consequence of this is that a more local hiring becomes possible for the companies.

LINK WITH THE BIM

LINK WITH THE BIM

Are Bim and Lean compatible?

More than compatible, BIM and Lean are complementary in many ways.

For the uninitiated, or as a reminder, BIM (Building Information Modeling) is a set of tools designed to carry out a construction site in parallel with a collaborative 3D computer model.

BIM is not a tool, but an methodology, a set of processes: this is why we will see 3 approaches to apply Lean to BIM.

LINK WITH THE BIM

The projects using BIM methodology so far presented additional cost generating issues such as:

- Necessary remodeling

- Too many reworkings

- Unstructured delivery

BIM, like Lean, is designed to reduce design and production problems.

LINK WITH THE BIM

A first approach is to perform a mapping of the "Workflows" of the following subjects:

- 3D BIM model production

- 3D BIM design coordination

- Production of a complete site model

- Standard BIM operation process

If you have no idea what I'm talking about, let's just say that it aims at modeling the processes for each deliverable.

LINK WITH THE BIM

Once your workflows are "mapped" (represented), waste analysis and identification can be performed.

Here are some examples of BIM-specific waste:

- **Skills:** contractors are not evaluated to use the software before the project.

- **Design:** Poor coordination, undefined information requirements and quality standards.

- **Meetings:** Unstructured coordination sessions, undefined or non-existent processes.

LINK WITH THE BIM

The identification and elimination of waste is the first step towards process improvement.

To anticipate them and allow the project to run smoothly, here is the next step:

- Develop standard workflows

- Define BIM deliverables during meetings

- Evaluate the BIM capacity of the designers in the pre-construction phase.

LINK WITH THE BIM

The second approach, the second Lean tool used, is called **SIPOC**. It is about representing the processes in a different way from the workflows presented previously.

S	I	P	O	C
Supplier	Input	Process	Output	Customer
Architect	Methods	Layout of plans	Archi Model	Modellers
...	...	...	...	...

For each step of the process, this table is made: we define a supplier, its prerequisites, its action, its output and to whom the deliverable is intended. The terms "supplier" and "customer" refer to internal actors of the BIM process.

LINK WITH THE BIM

To approach BIM with a Lean eye in a third way, 6 value losses were identified:

- Wrong modeling scope: too detailed or too generalized...

- Over-modeling or duplicate modeling: too much information, or information received at the wrong time.

- Waiting for deliverables, decisions, approvals...

It's funny, we find the same communication problems in design and project management as on site!

LINK WITH THE BIM

What other wastes can we work on?

Other impairments include:

- Sequencing: who has priority? For example, two pipes are passing at the same place, so one of them has to deviate: who pays and "lets the other pipe through"?

- Information may be missing in the model

- Re-modeling

LINK WITH THE BIM

To limit this waste as much as possible, the solution proposed by Lean is the following:

1) Make a strategic plan with a 12/18 month vision, with few milestones

2) Define the sequence and activity transfers on a 2 to 4 month vision

3) Prepare the program of deliverables on a 6-week vision.

This planning model, which aims at going for finer and finer details, is not unrelated to the LPS®; it is simply applied in design, not in production!

LINK WITH THE BIM

Since Lean or BIM work sites are not very common until now, the cases to be studied remain quite rare.

However, a rather revealing case study observed two groups conducting a BIM project, with or without Lean applied to the design.

The results are appealing:

- The group that did not apply Lean encountered 235 problems during design.

- The group that used LPS® collaborative planning encountered only 63 problems, 73% less than the first group!

CONCLUSION

CONCLUSION

Lean Construction is developing strongly today, as companies are increasingly aware of the importance of continuous improvement for the sustainability of a structure.

To begin with, make some initial analyses internally and try to launch the first 5S type projects to get the ball rolling; being accompanied by an external organization often helps to take a step back and give a boost to these projects.